Chapter 3
Multiply 2-Digit Numbers

Made in the United States
Text printed on 100% recycled paper

Houghton Mifflin Harcourt

Copyright © 2015 by Houghton Mifflin Harcourt Publishing Company

All rights reserved. No part of this work may be reproduced or transmitted in any form or by any means, electronic or mechanical, including photocopying or recording, or by any information storage and retrieval system, without the prior written permission of the copyright owner unless such copying is expressly permitted by federal copyright law. Requests for permission to make copies of any part of the work should be addressed to Houghton Mifflin Harcourt Publishing Company, Attn: Contracts, Copyrights, and Licensing, 9400 Southpark Center Loop, Orlando, Florida 32819-8647.

Common Core State Standards © Copyright 2010. National Governors Association Center for Best Practices and Council of Chief State School Officers. All rights reserved.

This product is not sponsored or endorsed by the Common Core State Standards Initiative of the National Governors Association Center for Best Practices and the Council of Chief State School Officers.

Printed in the U.S.A.

ISBN 978-0-544-34220-0

16 0928 22 21 20 19 18

4500743017   C D E F G

If you have received these materials as examination copies free of charge, Houghton Mifflin Harcourt Publishing Company retains title to the materials and they may not be resold. Resale of examination copies is strictly prohibited.

Possession of this publication in print format does not entitle users to convert this publication, or any portion of it, into electronic format.

Dear Students and Families,

Welcome to **Go Math!**, Grade 4! In this exciting mathematics program, there are hands-on activities to do and real-world problems to solve. Best of all, you will write your ideas and answers right in your book. In **Go Math!**, writing and drawing on the pages helps you think deeply about what you are learning, and you will really understand math!

By the way, all of the pages in your **Go Math!** book are made using recycled paper. We wanted you to know that you can Go Green with **Go Math!**

Sincerely,

The Authors

Made in the United States
Text printed on 100% recycled paper

GO MATH!

Authors

Juli K. Dixon, Ph.D.
Professor, Mathematics Education
University of Central Florida
Orlando, Florida

Edward B. Burger, Ph.D.
President, Southwestern University
Georgetown, Texas

Steven J. Leinwand
Principal Research Analyst
American Institutes for
   Research (AIR)
Washington, D.C.

Contributor

Rena Petrello
Professor, Mathematics
Moorpark College
Moorpark, California

Matthew R. Larson, Ph.D.
K-12 Curriculum Specialist for
   Mathematics
Lincoln Public Schools
Lincoln, Nebraska

Martha E. Sandoval-Martinez
Math Instructor
El Camino College
Torrance, California

English Language Learners Consultant

Elizabeth Jiménez
CEO, GEMAS Consulting
Professional Expert on English
   Learner Education
Bilingual Education and
   Dual Language
Pomona, California

Place Value and Operations with Whole Numbers

 Critical Area Developing understanding and fluency with multi-digit multiplication, and developing understanding of dividing to find quotients involving multi-digit dividends

 Multiply 2-Digit Numbers   143

COMMON CORE STATE STANDARDS

4.OA Operations and Algebraic Thinking
Cluster A Use the four operations with whole numbers to solve problems.
4.OA.A.3

4.NBT Number and Operations in Base Ten
Cluster B Use place value understanding and properties of operations to perform multi-digit arithmetic.
4.NBT.B.5

✓ Show What You Know **143**
   Vocabulary Builder **144**
   Chapter Vocabulary Cards
   Vocabulary Game **144A**

1 Multiply by Tens **145**
   Practice and Homework

2 Estimate Products **151**
   Practice and Homework

3 Investigate • Area Models and Partial Products **157**
   Practice and Homework

4 Multiply Using Partial Products **163**
   Practice and Homework

✓ Mid-Chapter Checkpoint **169**

Critical Area

GO DIGITAL

Go online! Your math lessons are interactive. Use *i*Tools, Animated Math Models, the Multimedia *e*Glossary, and more.

Chapter 3 Overview

In this chapter, you will explore and discover answers to the following **Essential Questions**:

- What strategies can you use to multiply 2-digit numbers?
- How can you use place value to multiply 2-digit numbers?
- How can you choose the best method to multiply 2-digit numbers?

Personal Math Trainer
Online Assessment and Intervention

5 Multiply with Regrouping **171**
Practice and Homework

6 Choose a Multiplication Method **177**
Practice and Homework

7 Problem Solving • Multiply 2-Digit Numbers **183**
Practice and Homework

✓ Chapter 3 Review/Test **189**

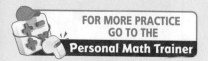

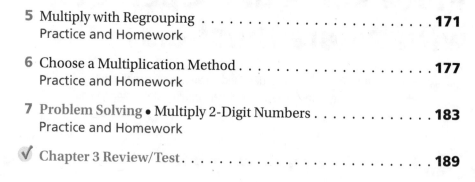

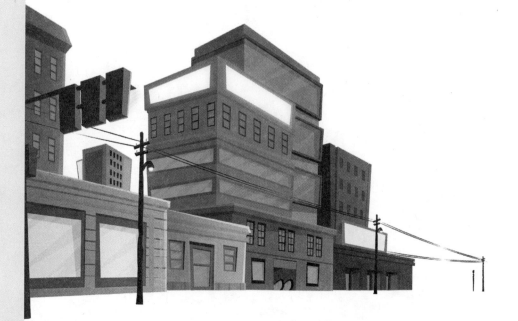

Chapter 3: Multiply 2-Digit Numbers

Show What You Know

Personal Math Trainer
Online Assessment and Intervention

Check your understanding of important skills.

Name _____

▶ **Practice Multiplication Facts** Find the product. (3.OA.C.7)

1. $8 \times 7 =$ _____
   $7 \times 8 =$ _____

2. $3 \times (2 \times 4) =$ _____
   $(3 \times 2) \times 4 =$ _____

▶ **2-Digit by 1-Digit Multiplication** Find the product. (4.NBT.B.5)

3. 28
   × 3

4. 56
   × 6

5. 71
   × 5

6. 69
   × 8

7. 36
   × 4

▶ **Multiply by 1-Digit Numbers** Find the product. (4.NBT.B.5)

8. 72
   × 4

9. 456
   × 5

10. 804
    × 7

11. 1,341
    × 9

12. 65
    × 6

13. 392
    × 8

14. 1,478
    × 3

15. $1,627
    × 2

16. 584
    × 7

17. 2,837
    × 4

Math in the Real World

Yellowstone National Park, which is located in Wyoming, Montana, and Idaho, was America's first National Park. The park has over 500 geysers. Grand Geyser erupts about every 8 hours.
Based on this estimate, how many times would you see this geyser erupt if you could watch it for 1 year? There are 24 hours in a day and 365 days in a year.

Chapter 3 143

Vocabulary Builder

▶ **Visualize It**

Complete the H-diagram using the words with a ✓.

Multiplication Words	Estimation Words

Review Words

Associative Property of Multiplication
Commutative Property of Multiplication
✓ estimate
✓ factor
✓ partial product
✓ place value
✓ product
regroup
✓ round

Preview Words

✓ compatible numbers

▶ **Understand Vocabulary**

Draw a line to match each word or phrase with its definition.

Word

1. Commutative Property of Multiplication

2. estimate

3. compatible numbers

4. factor

5. regroup

Definition

- A number that is multiplied by another number to find a product

- To exchange amounts of equal value to rename a number

- To find an answer that is close to the exact amount

- Numbers that are easy to compute mentally

- The property that states when the order of two factors is changed, the product is the same.

144
- Interactive Student Edition
- Multimedia eGlossary

Chapter 3 Vocabulary

Associative Property of Multiplication propiedad asociativa de la multiplicación 4	**Commutative Property of Multiplication** Propiedad conmutativa de la multiplicación 13
compatible numbers números compatibles 15	**estimate (*verb*)** estimar 30
factor factor 33	**partial product** producto parcial 61
place value valor posicional 68	**regroup** reagrupar 78

The property that states that when the order of two factors is changed, the product is the same

Example: $3 \times 5 = 5 \times 3$

The property that states that you can group factors in different ways and still get the same product

Example: $3 \times (4 \times 2) = (3 \times 4) \times 2$

To find an answer that is close to the exact amount

Numbers that are easy to compute mentally

Example: Estimate. $176 \div 8$

160 divides easily by 8

compatible number

A method of multiplying in which the ones, tens, hundreds, and so on are multiplied separately and then the products are added together

```
   182
 ×   6
   600
   480  ← Partial products
 +  12
 1,092
```

A number that is multiplied by another number to find a product

Example: $4 \times 5 = 20$

factor   factor

To exchange amounts of equal value to rename a number

Example: $5 + 8 = 13$ ones or 1 ten 3 ones

The value of a digit in a number, based on the location of the digit

Going Places with GO MATH! words

Matchup

For 3 to 4 players

Materials
- 1 set of word cards

How to Play

1. Put the cards face-down in rows. Take turns to play.
2. Choose two cards and turn them face-up.
   - If the cards show a word and its meaning, it's a match. Keep the pair and take another turn.
   - If the cards do not match, turn them back over.
3. The game is over when all cards have been matched. The players count their pairs. The player with the most pairs wins.

Word Box
Associative Property of Multiplication
Commutative Property of Multiplication
compatible numbers
estimate
factor
partial product
place value
regroup

Journal

The Write Way

Reflect

Choose one idea. Write about it.

- Do 36 × 29 and 29 × 36 represent the same product? Explain why or why not.
- Explain in your words what the Associative Property of Multiplication means.
- A reader of your math advice column writes, "I can't remember what compatible numbers are or how to use them." Write a letter that helps your reader with this problem.

Name _____

Multiply by Tens

Essential Question What strategies can you use to multiply by tens?

Lesson 3.1

 Number and Operations in Base Ten—4.NBT.B.5 Also 4.NBT.A.1
MATHEMATICAL PRACTICES
MP2, MP4, MP7

Unlock the Problem

Animation for a computer-drawn cartoon requires about 20 frames per second. How many frames would need to be drawn for a 30-second cartoon?

- The phrase "20 frames per second" means 20 frames are needed for each second of animation. How does this help you know what operation to use?

_____
_____

One Way Use place value.

Multiply. 30×20

You can think of 20 as 2 tens.

$30 \times 20 = 30 \times$ _____ tens

$ =$ _____ tens

$ = 600$

Another Way Use the Associative Property.

You can think of 20 as 2×10.

$30 \times 20 = 30 \times (2 \times 10)$

$ = (30 \times 2) \times 10$

$ =$ _____ $\times$ _____

$ =$ _____

So, _____ frames would need to be drawn.

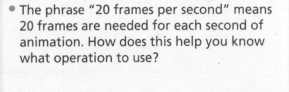

Remember
The Associative Property states that you can group factors in different ways and get the same product. Use parentheses to group the factors you multiply first.

 MATHEMATICAL PRACTICES 7
Look for Structure How can you use place value to tell why $60 \times 10 = 600$?

- Compare the number of zeros in each factor to the number of zeros in the product. What do you notice?

_____
_____

Chapter 3 145

Other Ways

A Use a number line and a pattern to multiply 15 × 20.

Draw jumps to show the product.

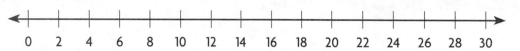

15 × 2 = _____

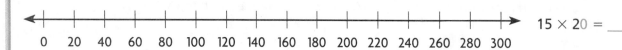

15 × 20 = _____

B Use mental math to find 14 × 30.

Use the halving-and-doubling strategy.

STEP 1 Find half of 14 to make the problem simpler.	**STEP 2** Multiply.	**STEP 3** Double 210.
Think: To find half of a number, divide by 2.		Think: To double a number, multiply by 2.
14 ÷ 2 = _____	7 × 30 = _____	2 × 210 = _____

So, 14 × 30 = 420.

Try This! Multiply.

Use mental math to find 12 × 40.	Use place value to find 12 × 40.

Share and Show

1. Find 20 × 27. Tell which method you chose. Explain what happens in each step.

_____

_____

Name _____

Choose a method. Then find the product.

2. 10 × 12

3. 20 × 20

✓4. 40 × 24

✓5. 11 × 60

Identify Relationships
How can you use 30 × 10 = 300 to find 30 × 12?

On Your Own

Choose a method. Then find the product.

6. 70 × 55

7. 17 × 30

8. 30 × 60

9. 12 × 90

 Reason Quantitatively Algebra Find the unknown digit in the number.

10. 64 × 40 = 2,56▪

11. 29 × 50 = 1,⬟50

12. 3◆ × 47 = 1,410

▪ = _____

⬟ = _____

◆ = _____

13.  Caroline packs 12 jars of jam in a box. She has 40 boxes. She has 542 jars of jam. How many jars of jam will she have left when all the boxes are full?

14.  Alison is preparing for a math contest. Each day, she works on multiplication problems for 20 minutes and division problems for 10 minutes. How many minutes does Alison practice multiplication and division problems in 15 days?

Chapter 3 • Lesson 1   147

Problem Solving • Applications

Use the table for 15–16.

15. **MATHEMATICAL PRACTICE 4** Use Graphs How many frames did it take to produce 50 seconds of *Pinocchio*?

_____

16. **GO DEEPER** Are there fewer frames in 10 seconds of *The Flintstones* or in 14 seconds of *The Enchanted Drawing?* What is the difference in the number of frames?

_____

Animated Productions

Title	Date Released	Frames per Second
The Enchanted Drawing©	1900	20
Little Nemo©	1911	16
Snow White and the Seven Dwarfs©	1937	24
Pinocchio©	1940	19
The Flintstones™	1960–1966	24

17. **THINK SMARTER** The product of my number and twice my number is 128. What is half my number? Explain how you solved the problem.

_____

_____

18. **THINK SMARTER** Tanya says that the product of a multiple of ten and a multiple of ten will always have only one zero. Is she correct? Explain.

_____

_____

WRITE Math • Show Your Work

19. **THINK SMARTER** For numbers 19a–19e, select Yes or No to tell whether the answer is correct.

19a. 28 × 10 = 280   ○ Yes   ○ No

19b. 15 × 20 = 300   ○ Yes   ○ No

19c. 17 × 10 = 17   ○ Yes   ○ No

19d. 80 × 10 = 800   ○ Yes   ○ No

19e. 16 × 30 = 1,800   ○ Yes   ○ No

Name _____

Multiply by Tens

**Practice and Homework
Lesson 3.1**

COMMON CORE STANDARD—4.NBT.B.5
Use place value understanding and properties of operations to perform multi-digit arithmetic.

Choose a method. Then find the product.

1. 16 × 60

   Use the halving-and-doubling strategy.

   Find half of 16: 16 ÷ 2 = 8.

   Multiply 60 by this number: 8 × 60 = 480

   Double this result: 2 × 480 = 960

   _____960_____

2. 80 × 22

3. 30 × 52

4. 60 × 20

_____   _____   _____

Problem Solving

5. Kenny bought 20 packs of baseball cards. There are 12 cards in each pack. How many cards did Kenny buy?

6. The Hart family drove 10 hours to their vacation spot. They drove an average of 48 miles each hour. How many miles did they drive?

_____   _____

7. **WRITE** ▸*Math* Write the steps for how to use a number line to multiply a 2-digit number by 20. Give an example.

_____

_____

_____

_____

Chapter 3 149

Lesson Check (4.NBT.B.5)

1. For the school play, 40 rows of chairs are set up. There are 22 chairs in each row. How many chairs are there?

2. At West School, there are 20 classrooms. Each classroom has 20 students. How many students are at West School?

Spiral Review (4.OA.A.1, 4.OA.A.2, 4.OA.A.3, 4.NBT.B.4)

3. Alex has 48 stickers. This is 6 times the number of stickers Max has. How many stickers does Max have?

4. Ali's dog weighs 8 times as much as her cat. Together, the two pets weigh 54 pounds. How much does Ali's dog weigh?

5. Allison has 3 containers with 25 crayons in each. She also has 4 boxes of markers with 12 markers in each box. She gives 10 crayons to a friend. How many crayons and markers does Allison have now?

6. The state of Utah covers 82,144 square miles. The state of Montana covers 145,552 square miles. What is the total area of the two states?

Name _____

Estimate Products

Essential Question What strategies can you use to estimate products?

Lesson 3.2

Common Core — Number and Operations in Base Ten—4.NBT.B.5 Also 4.NBT.A.3
MATHEMATICAL PRACTICES
MP1, MP2, MP6

Unlock the Problem

On average, the Smith family opens the door of their refrigerator 32 times each day. There are 31 days in May. About how many times is the refrigerator door opened in May?

• Underline any information you will need.

One Way Use rounding and mental math.

Estimate. 31 × 32

STEP 1 Round each factor.

31 × 32
↓   ↓
30 × 30

STEP 2 Use mental math.

3 × 3 = 9 ← basic fact

30 × 30 = _____

Math Talk

MATHEMATICAL PRACTICES 6

Compare Is the exact product greater than or less than 900? Explain.

So, the Smith family opens the refrigerator door about 900 times during the month of May.

1. On average, a refrigerator door is opened 38 times each day. About how many fewer times in May is the Smith family's refrigerator door opened than the average refrigerator door?

Show your work.

Chapter 3   151

All 24 light bulbs in the Park family's home are CFL light bulbs. Each CFL light bulb uses 28 watts to produce light. About how many watts will the light bulbs use when turned on all at the same time?

Another Way Use mental math and compatible numbers.

Compatible numbers are numbers that are easy to compute mentally.

Estimate. 24 × 28

STEP 1 Use compatible numbers.

24 × 28
↓   ↓
25 × 30   Think: 25 × 3 = 75

So, about 750 watts are used.

STEP 2 Use mental math.

25 × 3 = 75

25 × 30 = _____

Try This! Estimate 26 × $79.

A Round to the nearest ten

26   × $79
↓   ↓
____ × ____ = ____

26 × $79 is about _____.

B Compatible numbers

26   × $79
↓   ↓
25   × $80 = _____

Think: How can you use 25 × 4 = 100 to help find 25 × 8?

26 × $79 is about _____.

2. Explain why $2,400 and $2,000 are both reasonable estimates.

_____

_____

3. In what situation might you choose to find an estimate rather than an exact answer?

_____

_____

Share and Show

1. To estimate the product of 62 and 28 by rounding, how would you round the factors? What would the estimated product be?

_____

Name _____

Estimate the product. Choose a method.

2. 96 × 34

3. 47 × $39

4. 78 × 72

Math Talk MATHEMATICAL PRACTICES ①
Describe how you know if an estimated product will be greater than or less than the exact answer.

On Your Own

Estimate the product. Choose a method.

5. 41 × 78

6. 51 × 73

7. 34 × 80

Practice: Copy and Solve Estimate the product. Choose a method.

8. 61 × 31

9. 52 × 68

10. 26 × 44

11. 57 × $69

THINK SMARTER Find two possible factors for the estimated product.

12. 2,800

13. 8,100

14. 5,600

15. 2,400

16. **GO DEEPER** Mr. Parker jogs for 35 minutes each day. He jogs 5 days in week 1, 6 days in week 2, and 7 days in week 3. About how many minutes does he jog?

17. **GO DEEPER** There are 48 beads in a package. Candice bought 4 packages of blue, 9 packages of gold, 6 packages of red, and 2 packages of silver beads. About how many beads did Candice buy?

Chapter 3 • Lesson 2 153

MATHEMATICAL PRACTICES • COMMUNICATE • PERSEVERE • CONSTRUCT ARGUMENTS

Problem Solving • Applications

18. **GO DEEPER** On average, a refrigerator door is opened 38 times each day. Len has two refrigerators in his house. Based on this average, about how many times in a 3-week period are the refrigerator doors opened?

_____

19. The cost to run a refrigerator is about $57 each year. About how much will it have cost to run by the time it is 15 years old?

_____

20. **THINK SMARTER** If Mel opens his refrigerator door 36 times every day, about how many times will it be opened in April? Will the exact answer be more than or less than the estimate? Explain.

_____

_____

_____

21. **MATHEMATICAL PRACTICE 2** Represent a Problem What question could you write for this answer? The estimated product of two numbers, that are not multiples of ten, is 2,800.

WRITE *Math* • Show Your Work

_____

_____

22. **THINK SMARTER** Which is a reasonable estimate for the product? Write the estimate. An estimate may be used more than once.

| 30 × 20 | 25 × 50 | 20 × 20 |

26 × 48 _____   28 × 21 _____

21 × 22 _____   51 × 26 _____

154

Name _____

Estimate Products

**Practice and Homework
Lesson 3.2**

COMMON CORE STANDARD—4.NBT.B.5
Use place value understanding and properties of operations to perform multi-digit arithmetic.

Estimate the product. Choose a method.

1. 38 × 21

   38 × 21
   ↓   ↓
   40 × 20

   ___800___

2. 63 × 19

   _____

3. 27 × $42

   _____

4. 73 × 67

   _____

5. 37 × $44

   _____

6. 45 × 22

   _____

Problem Solving

7. A dime has a diameter of about 18 millimeters. About how many millimeters long would a row of 34 dimes be?

   _____

8. A half-dollar has a diameter of about 31 millimeters. About how many millimeters long would a row of 56 half-dollars be?

   _____

9. **WRITE** Math Describe a real-life multiplication situation for which an estimate makes sense. Explain why it makes sense.

   _____

   _____

Chapter 3   155

Lesson Check (4.NBT.B.5)

1. What is a reasonable estimate for the product of 43 × 68?

   _____

   _____

2. Marissa burns 93 calories each time she plays fetch with her dog. She plays fetch with her dog once a day. About how many calories will Marissa burn playing fetch with her dog in 28 days?

   _____

   _____

Spiral Review (4.NBT.A.1, 4.NBT.A.3, 4.NBT.B.5)

3. Use the model to find 3 × 126.

   ```
         100      20  6
      ┌─────────┬────┬──┐
    3 │         │    │  │
      └─────────┴────┴──┘
   ```

4. A store sold a certain brand of jeans for $38. One day, the store sold 6 pairs of jeans of that brand. How much did the 6 pairs of jeans cost?

   _____

   _____

5. The Gateway Arch in St. Louis, Missouri, weighs about 20,000 tons. Write an amount that could be the exact number of tons the Arch weighs.

   _____

   _____

6. What is another name for 23 ten thousands?

   _____

   _____

156

Name _____

Area Models and Partial Products

Essential Question How can you use area models and partial products to multiply 2-digit numbers?

Lesson 3.3

Common Core — Number and Operations in Base Ten—4.NBT.B.5
MATHEMATICAL PRACTICES
MP2, MP4, MP6

Investigate

Materials ■ color pencils

How can you use a model to break apart factors and make them easier to multiply?

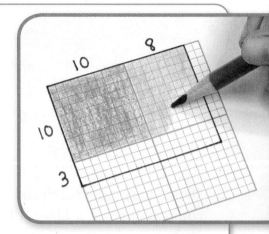

A. Outline a rectangle on the grid to model 13 × 18. Break apart the model into smaller rectangles to show factors broken into tens and ones. Label and shade the smaller rectangles. Use the colors below.

B. Find the product of each smaller rectangle. Then, find the sum of the partial products. Record your answers.

▢ = 10 × 10
▢ = 10 × 8
▢ = 3 × 10
▢ = 3 × 8

100 + ___ + ___ + ___ = ___

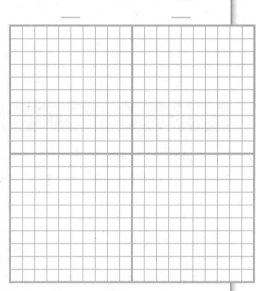

C. Draw the model again. Break apart the whole model to show factors different from those shown the first time. Label and shade the four smaller rectangles and find their products. Record the sum of the partial products to represent the product of the whole model.

___ + ___ + ___ + ___ = ___

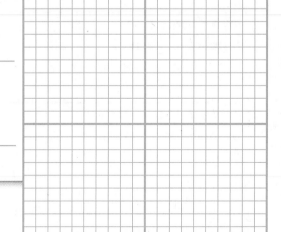

Chapter 3 157

Draw Conclusions

1. Explain how you found the total number of squares in the whole model.

   _____

   _____

2. Compare the two models and their products. What can you conclude? Explain.

   _____

   _____

   _____

3. To find the product of 10 and 33, which is the easier computation, $(10 \times 11) + (10 \times 11) + (10 \times 11)$ or $(10 \times 30) + (10 \times 3)$? Explain.

   _____

   _____

   _____

Make Connections

You can draw a simple diagram to model and break apart factors to find a product. Find 15×24.

Remember

24 is 2 tens 4 ones.

STEP 1 Draw a model to show 15×24. Break apart the factors into tens and ones to show the partial products.

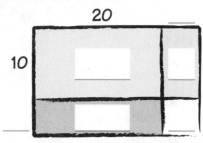

STEP 2 Write the product for each of the smaller rectangles.

(10 × 2 tens) (10 × 4 ones) (5 × 2 tens) (5 × 4 ones)
(10 × 20)   (10 × 4)   (5 × 20)   (5 × 4)

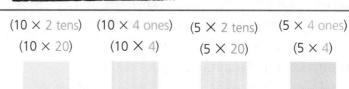

STEP 3 Add to find the product for the whole model.

 +  = _____

So, $15 \times 24 = 360$.

The model shows four parts. Each part represents a partial product. The partial products are 200, 40, 100, and 20.

Math Talk

MATHEMATICAL PRACTICES ②

Use Reasoning How does breaking apart the factors into tens and ones make finding the product easier?

Name _____

Share and Show

Find the product.

1. 16 × 19 = _____

2. 18 × 26 = _____

3. 27 × 39 = _____

**Draw a model to represent the product.
Then record the product.**

4. 14 × 16 = _____

5. 23 × 25 = _____

Problem Solving • Applications

6. **MATHEMATICAL PRACTICE 6** Explain how modeling partial products can be used to find the products of greater numbers.

_____

_____

7. **GO DEEPER** Emma bought 16 packages of rolls for a party. There were 12 rolls in a package. After the party there were 8 rolls left over. How many rolls were eaten? Explain.

_____

_____

MATHEMATICAL PRACTICES ANALYZE • LOOK FOR STRUCTURE • PRECISION

Sense or Nonsense?

8. **THINK SMARTER** Jamal and Kim used different ways to solve 12 × 15 by using partial products. Whose answer makes sense? Whose answer is nonsense? Explain your reasoning.

Jamal's Work

	10	5
10	100	50
2	20	10

100 + 20 + 10 = 130

_____
_____
_____

Kim's Work

	10	5
10	100	50
2	20	10

120 + 60 = 180

_____
_____
_____

a. For the answer that is nonsense, write an answer that makes sense.

_____

b. Look at Kim's method. Can you think of another way Kim could use the model to find the product? Explain.

_____
_____
_____

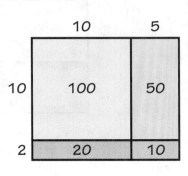

9. **THINK SMARTER** Look at the model in 8b. How would the partial products change if the product was 22 × 15? Explain why you think the products changed.

_____
_____

Name _____

Area Models and Partial Products

**Practice and Homework
Lesson 3.3**

COMMON CORE STANDARD—4.NBT.B.5
Use place value understanding and properties of operations to perform multi-digit arithmetic.

**Draw a model to represent the product.
Then record the product.**

1. 13 × 42

	40	2
10	400	20
3	120	6

400 + 20 + 120 + 6 = __546__

2. 18 × 34

_____

3. 22 × 26

_____

Problem Solving

4. Sebastian made the following model to find the product 17 × 24.

	20	4
10	200	40
7	14	28

200 + 40 + 14 + 28 = 282

Is his model correct? **Explain**.

_____

_____

5. Each student in Ms. Sike's kindergarten class has a box of crayons. Each box has 36 crayons. If there are 18 students in Ms. Sike's class, how many crayons are there?

_____

6. **WRITE** *Math* Describe how to model 2-digit by 2-digit multiplication using an area model.

_____

_____

Chapter 3   161

Lesson Check (4.NBT.B.5)

1. What product does the model below represent?

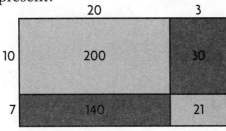

_____

2. What product does the model below represent?

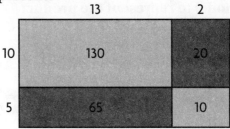

_____

Spiral Review (4.OA.A.3, 4.NBT.B.5)

3. Mariah builds a tabletop using square tiles. There are 12 rows of tiles and 30 tiles in each row. How many tiles does Mariah use?

_____

4. Trevor bakes 8 batches of biscuits, with 14 biscuits in each batch. He sets aside 4 biscuits from each batch for a bake sale and puts the rest in a container. How many biscuits does Trevor put in the container?

_____

5. Li feeds her dog 3 cups of food each day. About how many cups of food does her dog eat in 28 days?

_____

6. Find the product of $20 \times 9 \times 5$. Tell which property you used.

_____

Lesson 3.4

Name _____

Multiply Using Partial Products

Essential Question How can you use place value and partial products to multiply 2-digit numbers?

 Number and Operations in Base Ten—4.NBT.B.5
MATHEMATICAL PRACTICES
MP1, MP2, MP4, MP8

Unlock the Problem

CONNECT You know how to break apart a model to find partial products. How can you use what you know to find and record a product?

Multiply. 34 × 57   Estimate. 30 × 60 = _____

SHADE THE MODEL | **THINK AND RECORD**

STEP 1

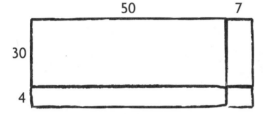

```
   57
 × 34
 ____
```
← Multiply the tens by the tens.
  30 × 5 tens = 150 tens

STEP 2

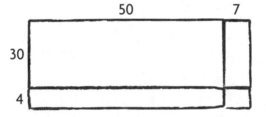

```
    57
  × 34
  ____
 1,500
```
← Multiply the ones by the tens.
  30 × 7 ones = 210 ones

STEP 3

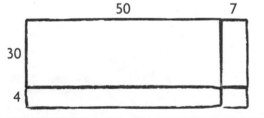

```
    57
  × 34
  ____
 1,500
   210
```
← Multiply the tens by the ones.
  4 × 5 tens = 20 tens

STEP 4

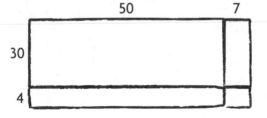

```
    57
  × 34
  ____
 1,500
   210
   200
 +
```
← Multiply the ones by the ones.
  4 × 7 ones = 28 ones
← Add the partial products.

So, 34 × 57 = 1,938. Since 1,938 is close to the estimate of 1,800, it is reasonable.

Math Talk

MATHEMATICAL PRACTICES 8

Use Repeated Reasoning You can write 10 × 4 ones = 40 ones as 10 × 4 = 40. What is another way to write 10 × 3 tens = 30 tens?

Chapter 3 163

Example

The apples from each tree in an orchard can fill 23 bushel baskets. If 1 row of the orchard has 48 trees, how many baskets of apples can be filled?

Multiply. 48 × 23   **Estimate.** 50 × 20 = _____

THINK	RECORD

STEP 1

Multiply the tens by the tens.

```
   23
 × 48
 ____
```
← 40 × _____ tens = _____ tens

STEP 2

Multiply the ones by the tens.

```
   23
 × 48
 ____
  800
```
← 40 × _____ ones = _____ ones

STEP 3

Multiply the tens by the ones.

```
   23
 × 48
 ____
  800
  120
```
← 8 × _____ tens = _____ tens

STEP 4

Multiply the ones by the ones. Then add the partial products.

```
   23
 × 48
 ____
  800
  120
  160
+ ____
```
← 8 × _____ ones = _____ ones

So, 1,104 baskets can be filled.

Math Talk

MATHEMATICAL PRACTICES ①

Evaluate Reasonableness How do you know your answer is reasonable?

Share and Show

1. Find 24 × 34.

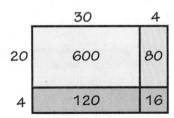

164

Name _____

Record the product.

2. 12
   × 12

3. 31
   × 24

4. 25
   × 43

5. 37
   × 26

Math Talk

MATHEMATICAL PRACTICES 4

Model Mathematics How would you model and record 74 × 25?

On Your Own

Record the product.

6. 54
   × 15

7. 87
   × 16

8. 62
   × 56

9. 49
   × 63

Practice: Copy and Solve Record the product.

10. 38 × 47
11. 46 × 27
12. 72 × 53
13. 98 × 69

14. 53 × 68
15. 76 × 84
16. 92 × 48
17. 37 × 79

MATHEMATICAL PRACTICE 2 Reason Abstractly **Algebra** Find the unknown digits. Complete the problem.

18. ☐ 6
    ×   ☐ 4
    1,400
      120
      280
    + 24
    ─────

19. ☐ 2
    × ☐ 7
    7,200
      180
      560
    + 14
    ─────

20. ☐ 6
    × ☐ 5
    1,500
      300
       90
    + 18
    ─────

21.   3 ☐
    × ☐ 8
      600
       80
      240
    + 32
    ─────

Chapter 3 • Lesson 4   165

Problem Solving • Applications

Use the picture graph for 22–24.

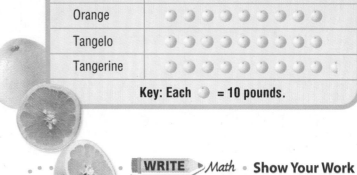

Pounds of Citrus Fruit per Box

Citrus Fruit	Weight per Box (in pounds)
Grapefruit	◐◐◐◐◐◐◐◖
Orange	◐◐◐◐◐◐
Tangelo	◐◐◐◐◐◐◐
Tangerine	◐◐◐◐◐◐◐◐◖

Key: Each ◐ = 10 pounds.

22. **Use Graphs** A fruit-packing warehouse is shipping 15 boxes of grapefruit to a store in Santa Rosa, California. What is the total weight of the shipment?

23. **GO DEEPER** How much less do 13 boxes of tangelos weigh than 18 boxes of tangerines?

24. What is the weight of 12 boxes of oranges?

25. **THINK SMARTER** Each person in the United States eats about 65 fresh apples each year. Based on this estimate, how many apples do 3 families of 4 eat each year?

26. **GO DEEPER** The product 26 × 93 is greater than 25 × 93. How much greater? Explain how you know without multiplying.

27. **THINK SMARTER** Margot wants to use partial products to find 22 × 17. Write the numbers in the boxes to show 22 × 17.

(☐ × ☐) + (☐ × ☐) + (☐ × ☐) + (☐ × ☐)

166

Name _____

Multiply Using Partial Products

**Practice and Homework
Lesson 3.4**

COMMON CORE STANDARD—4.NBT.B.5
Use place value understanding and properties of operations to perform multi-digit arithmetic.

Record the product.

1. 23
    × 79
    ─────
    1,400
      210
      180
    + 27
    ─────
    1,817

2. 56
    × 32

3. 87
    × 64

4. 33
    × 25

5. 94
    × 12

6. 51
    × 77

7. 69
    × 49

Problem Solving

8. Evelyn drinks 8 glasses of water a day, which is 56 glasses of water a week. How many glasses of water does she drink in a year? (1 year = 52 weeks)

   _____

9. Joe wants to use the Hiking Club's funds to purchase new walking sticks for each of its 19 members. The sticks cost $26 each. The club has $480. Is this enough money to buy each member a new walking stick? If not, how much more money is needed?

   _____

10. **WRITE** *Math* Explain why it works to break apart a number by place values to multiply.

    _____
    _____
    _____
    _____

Lesson Check (4.NBT.B.5)

1. A carnival snack booth made $76 selling popcorn in one day. It made 22 times as much selling cotton candy. How much money did the snack booth make selling cotton candy?

2. List the partial products of 42 × 28.

Spiral Review (4.OA.A.1, 4.OA.A.3, 4.NBT.B.5)

3. Last year, the city library collected 117 used books for its shelves. This year, it collected 3 times as many books. How many books did it collect this year?

4. Washington Elementary has 232 students. Washington High has 6 times as many students. How many students does Washington High have?

5. List the partial products of 35 × 7.

6. Shelby has ten $5 bills and thirteen $10 bills. How much money does Shelby have in all?

Name _____

Mid-Chapter Checkpoint

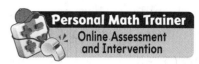

Concepts and Skills

1. Explain how to find 40 × 50 using mental math. (4.NBT.B.5)

   _____

   _____

2. What is the first step in estimating 56 × 27? (4.NBT.B.5)

   _____

Choose a method. Then find the product. (4.NBT.B.5)

3. 35 × 10 _____   4. 19 × 20 _____   5. 12 × 80 _____

6. 70 × 50 _____   7. 58 × 40 _____   8. 30 × 40 _____

9. 14 × 60 _____   10. 20 × 30 _____   11. 16 × 90 _____

Estimate the product. Choose a method. (4.NBT.B.5)

12. 81 × 38 _____   13. 16 × $59 _____   14. 43 × 25 _____

15. 76 × 45 _____   16. 65 × $79 _____   17. 92 × 38 _____

18. 37 × 31 _____   19. 26 × $59 _____   20. 54 × 26 _____

21. 52 × 87 _____   22. 39 × 27 _____   23. 63 × 58 _____

Chapter 3 169

24. Ms. Traynor's class is taking a field trip to the zoo. The trip will cost $26 for each student. There are 22 students in her class. What is a good estimate for the cost of the students' field trip? (4.NBT.B.5)

_____

25. Tito wrote the following on the board. What is the unknown number? (4.NBT.B.5)

$$50 \times 80 = 50 \times (8 \times 10)$$
$$= (50 \times 8) \times 10$$
$$= ? \times 10$$
$$= 4{,}000$$

_____

26. What are the partial products that result from multiplying 15×32? (4.NBT.B.5)

_____

27. **GO DEEPER** A city bus company sold 39 one-way tickets and 20 round-trip tickets from West Elmwood to East Elmwood. One-way tickets cost $14. Round trip tickets cost $25. How much money did the bus company collect? (4.NBT.B.5)

_____

Name _____

Multiply with Regrouping

Essential Question How can you use regrouping to multiply 2-digit numbers?

Lesson 3.5

 Number and Operations in Base Ten—4.NBT.B.5 Also 4.OA.A.3
MATHEMATICAL PRACTICES
MP2, MP7, MP8

Unlock the Problem

By 1914, Henry Ford had streamlined his assembly line to make a Model T Ford car in 93 minutes. How many minutes did it take to make 25 Model Ts?

 Use place value and regrouping.

Multiply. 93 × 25   **Estimate.** 90 × 30 = _____

▲ The first production Model T Ford was assembled on October 1, 1908.

	THINK	RECORD
STEP 1	• Think of 93 as 9 tens and 3 ones. • Multiply 25 by 3 ones.	$\overset{1}{2}5$ $\times93$ ☐ ← 3 × 25
STEP 2	• Multiply 25 by 9 tens.	$\overset{4}{\cancel{1}}$ 25 $\times93$ 75 ☐ ← 90 × 25
STEP 3	• Add the partial products.	$\overset{4}{\cancel{1}}$ 25 $\times93$ 75 $2{,}250$

So, 93 × 25 is 2,325. Since _____ is close to the estimate of _____, the answer is reasonable.

 Math Talk

MATHEMATICAL PRACTICES 8
Use Repeated Reasoning
Why do you get the same answer whether you multiply 93 × 25 or 25 × 93?

Chapter 3   171

Different Ways to Multiply You can use different ways to multiply and still get the correct answer. Shawn and Patty both solved 67 × 40 correctly, but they used different ways.

Look at Shawn's paper.

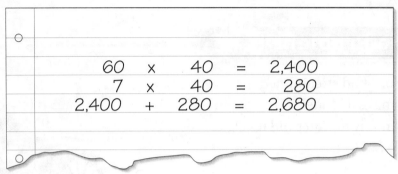

So, Shawn's answer is 67 × 40 = 2,680.

Look at Patty's paper.

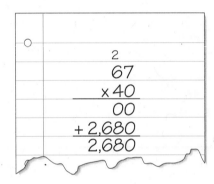

So, Patty also found 67 × 40 = 2,680.

1. What method did Shawn use to solve the problem?

_____

2. What method did Patty use to solve the problem?

_____

Share and Show

1. Look at the problem. Complete the sentences.

   Multiply ____ and ____ to get 0.

   Multiply ____ and ____ to get 1,620.

   Add the partial products.

   0 + 1,620 = _____

   $$\begin{array}{r} \overset{4}{27} \\ \times 60 \\ \hline 0 \\ +1{,}620 \\ \hline \end{array}$$

172

Name _____

Estimate. Then find the product.

2. Estimate: _____
   68
   × 53
   _____

3. Estimate: _____
   61
   × 54
   _____

4. Estimate: _____
   90
   × 27
   _____

Math Talk MATHEMATICAL PRACTICES ⑧

Generalize Why can you omit zeros of the first partial product when you multiply 20 × 34?

On Your Own

Estimate. Then find the product.

5. Estimate: _____
   30
   × 47
   _____

6. Estimate: _____
   78
   × 56
   _____

7. Estimate: _____
   27
   × 25
   _____

Practice: Copy and Solve Estimate. Then find the product.

8. 34 × 65
9. 42 × $13
10. 60 × 17
11. 62 × 45
12. 57 × $98

MATHEMATICAL PRACTICE ⑦ Look for a Pattern Algebra Write a rule for the pattern. Use your rule to find the unknown numbers.

13.

Hours	h	5	10	15	20	25
Minutes	m	300	600	900		

Rule: _____

14. **GO DEEPER** Owners of a summer camp are buying new cots for their cabins. There are 16 cabins. Each cabin needs 6 cots. Each cot costs $92. How much will the new cots cost?

15. **GO DEEPER** A theater has 28 rows of 38 seats downstairs and 14 rows of 26 seats upstairs. How many seats does the theater have?

Chapter 3 • Lesson 5 173

MATHEMATICAL PRACTICES COMMUNICATE • PERSEVERE • CONSTRUCT ARGUMENTS

Unlock the Problem

16. **THINK SMARTER** Machine A can label 11 bottles in 1 minute. Machine B can label 12 bottles in 1 minute. How many bottles can both machines label in 15 minutes?

a. What do you need to know? _____

b. What numbers will you use? _____

c. Tell why you might use more than one operation to solve the problem.

d. Solve the problem.

So, both machines can label _____ bottles in _____ minutes.

17. **MATHEMATICAL PRACTICE 1** Make Sense of Problems A toy company makes wooden blocks. A carton holds 85 blocks. How many blocks can 19 cartons hold?

18. **GO DEEPER** A company is packing cartons of candles. Each carton can hold 75 candles. So far, 50 cartons have been packed, but only 30 cartons have been loaded on a truck. How many more candles are left to load on the truck?

Personal Math Trainer

19. **THINK SMARTER+** Mr. Garcia's class raised money for a field trip to the zoo. There are 23 students in his class. The cost of the trip will be $17 for each student. What is the cost for all the students? Explain how you found your answer.

174

Multiply with Regrouping

Practice and Homework Lesson 3.5

COMMON CORE STANDARD—4.NBT.B.5
Use place value understanding and properties of operations to perform multi-digit arithmetic.

Estimate. Then find the product.

1. Estimate: __2,700__

   ```
     2
     1
     87
   × 32
   ─────
    174
   +2,610
   ─────
   2,784
   ```

   Think: 87 is close to 90 and 32 is close to 30.

   $90 \times 30 = 2,700$

2. Estimate: _____

   ```
     73
   × 28
   ```

3. Estimate: _____

   ```
     48
   × 38
   ```

4. Estimate: _____

   ```
     59
   × 52
   ```

Problem Solving

5. Baseballs come in cartons of 84 baseballs. A team orders 18 cartons of baseballs. How many baseballs does the team order?

   _____

6. There are 16 tables in the school lunch room. Each table can seat 22 students. How many students can be seated at lunch at one time?

   _____

7. **WRITE** *Math* Write about which method you prefer to use to multiply two 2-digit numbers—regrouping, partial products, or breaking apart a model. Explain why.

   _____

   _____

   _____

   _____

Lesson Check (4.NBT.B.5)

1. The art teacher has 48 boxes of crayons. There are 64 crayons in each box. How many crayons does the teacher have?

2. A basketball team scored an average of 52 points in each of 15 games. Based on the average, how many points did the team score in all?

Spiral Review (4.OA.A.1, 4.OA.A.2, 4.OA.A.3, 4.NBT.B.5)

3. One Saturday, an orchard sold 83 bags of apples. There are 27 apples in each bag. How many apples were sold?

4. Hannah has a grid of squares that has 12 rows with 15 squares in each row. She colors 5 rows of 8 squares in the middle of the grid blue. She colors the rest of the squares red. How many squares does Hannah color red?

5. Gabriella has 4 times as many erasers as Leona. Leona has 8 erasers. How many erasers does Gabriella have?

6. Phil has 3 times as many rocks as Peter. Together, they have 48 rocks. How many more rocks does Phil have than Peter?

Name _____

Choose a Multiplication Method

Essential Question How can you find and record products of two 2-digit numbers?

Lesson 3.6

Common Core — **Number and Operations in Base Ten—4.NBT.B.5**

MATHEMATICAL PRACTICES
MP6, MP7, MP8

 Unlock the Problem Real World

Did you know using math can help prevent you from getting a sunburn?

The time it takes to burn without sunscreen multiplied by the SPF, or sun protection factor, is the time you can stay in the sun safely with sunscreen.

If today's UV index is 8, Erin will burn in 15 minutes without sunscreen. If Erin puts on lotion with an SPF of 25, how long will she be protected?

- Underline the sentence that tells you how to find the answer.
- Circle the numbers you need to use. What operation will you use?
  _____

One Way Use partial products to find 15 × 25.

```
      25
    × 15
    ____
    ____  ← 10 × 2 tens = 20 tens
    ____  ← 10 × 5 ones = 50 ones
    ____  ← 5 × 2 tens = 10 tens
  + ____  ← 5 × 5 ones = 25 ones
    ____  ← Add.
```

▲ Sunscreen helps to prevent sunburn.

Draw a picture to check your work.

So, if Erin puts on lotion with an SPF of 25, she will be protected for 375 minutes.

 Math Talk

MATHEMATICAL PRACTICES 6

Explain how it was easier to find the product using partial products.

Chapter 3 177

Another Way Use regrouping to find 15 × 25.

Estimate. 20 × 20 = _____

STEP 1

Think of 15 as 1 ten 5 ones.
Multiply 25 by 5 ones, or 5.

$$\begin{array}{r} \overset{2}{2}5 \\ \times\ 15 \\ \hline \end{array}$$ ← 5 × 25

STEP 2

Multiply 25 by 1 ten, or 10.

$$\begin{array}{r} \overset{2}{2}5 \\ \times\ 15 \\ \hline 125 \\ \end{array}$$ ← 10 × 25

STEP 3

Add the partial products.

$$\begin{array}{r} \overset{2}{2}5 \\ \times\ 15 \\ \hline 125 \\ +\ 250 \\ \hline \end{array}$$

Try This! Multiply. 57 × $43

Estimate. 57 × $43	Use partial products.	Use regrouping.
	$43 × 57	$43 × 57

1. How do you know your answer is reasonable?

   _____

   _____

2. Look at the partial products and regrouping methods above. How are the partial products 2,000 and 150 related to 2,150?

   _____

   How are the partial products 280 and 21 related to 301?

   _____

178

Name _____

Share and Show

1. Find the product.

```
     5 4
  ×  2 9
```

Math Talk — MATHEMATICAL PRACTICES ⑧
Draw Conclusions Why do you begin with the ones place when you use the regrouping method to multiply?

Estimate. Then choose a method to find the product.

2. Estimate: _____
   36
   × 14

3. Estimate: _____
   63
   × 42

✓4. Estimate: _____
   84
   × 53

✓5. Estimate: _____
   71
   × 13

On Your Own

Practice: Copy and Solve Estimate. Find the product.

6. 29 × $82
7. 57 × 79
8. 80 × 27
9. 32 × $75

10. 55 × 48
11. 19 × $82
12. 25 × $25
13. 41 × 98

MATHEMATICAL PRACTICE ⑦ **Identify Relationships** **Algebra** Use mental math to find the number.

14. 30 × 14 = 420, so 30 × 15 = _____.

15. 25 × 12 = 300, so 25 × _____ = 350.

16. MATHEMATICAL PRACTICE ⑥ The town conservation manager bought 16 maple trees for $26 each. She paid with five $100 bills. How much change will the manager receive? **Explain**.

_____

_____

17. GO DEEPER Each of 25 students in Group A read for 45 minutes. Each of 21 students in Group B read for 48 minutes. Which group read for more minutes? Explain.

_____

Chapter 3 • Lesson 6 179

Common Core MATHEMATICAL PRACTICES ANALYZE • LOOK FOR STRUCTURE • PRECISION

Unlock the Problem

18. **THINK SMARTER** Martin collects stamps. He counted 48 pages in his collector's album. The first 20 pages each have 35 stamps in 5 rows. The rest of the pages each have 54 stamps. How many stamps does Martin have in his album?

a. What do you need to know? _____

_____

b. How will you use multiplication to find the number of stamps? _____

_____

c. Tell why you might use addition and subtraction to help solve the problem.

_____

_____

d. Show the steps to solve the problem.

e. Complete the sentences.

Martin has a total of _____ stamps on the first 20 pages.

There are _____ more pages after the first 20 pages in Martin's album.

There are _____ stamps on the rest of the pages.

There are _____ stamps in the album.

19. **THINK SMARTER** Select the expressions that have the same product as 35×17. Mark all that apply.

○ $(30 \times 10) + (30 \times 7) + (5 \times 10) + (5 \times 7)$

○ $(30 \times 17) + (5 \times 17)$

○ $(35 \times 30) + (35 \times 5) + (35 \times 10) + (35 \times 7)$

○ $(35 \times 10) + (35 \times 7)$

○ $(35 \times 10) + (30 \times 10) + (5 \times 10) + (5 \times 7)$

○ $(35 \times 30) + (35 \times 5)$

Name _____

Choose a Multiplication Method

**Practice and Homework
Lesson 3.6**

 COMMON CORE STANDARD—4.NBT.B.5
Use place value understanding and properties of operations to perform multi-digit arithmetic.

Estimate. Then choose a method to find the product.

1. Estimate: __1,200__

```
      31
   ×  43
      93
 + 1,240
   1,333
```

2. Estimate: _____

```
    67
 ×  85
```

3. Estimate: _____

```
    68
 ×  38
```

4. Estimate: _____

```
    95
 ×  17
```

5. Estimate: _____

```
    49
 ×  54
```

6. Estimate: _____

```
    91
 ×  26
```

7. Estimate: _____

```
    82
 ×  19
```

 Problem Solving

8. A movie theatre has 26 rows of seats. There are 18 seats in each row. How many seats are there?

   _____

9. Each class at Briarwood Elementary collected at least 54 cans of food during the food drive. If there are 29 classes in the school, what was the least number of cans collected?

   _____

10. **WRITE** ▸ *Math* How is multiplication using partial products different from multiplication using regrouping? How are they similar?

    _____

    _____

    _____

Lesson Check (4.NBT.B.5)

1. A choir needs new robes for each of its 46 singers. Each robe costs $32. What will be the total cost for all 46 robes?

2. A wall on the side of a building is made up of 52 rows of bricks with 44 bricks in each row. How many bricks make up the wall?

Spiral Review (4.NBT.B.4, 4.NBT.B.5)

3. Write an expression that shows how to multiply 4 × 362 using place value and expanded form.

4. Use the model below. What is the product 4 × 492?

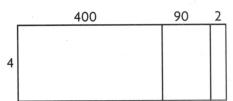

5. What is the sum 13,094 + 259,728?

6. During the 2008–2009 season, there were 801,372 people who attended the home hockey games in Philadelphia. There were 609,907 people who attended the home hockey games in Phoenix. How much greater was the home attendance in Philadelphia than in Phoenix that season?

Name _____

Problem Solving • Multiply 2-Digit Numbers

Essential Question How can you use the strategy *draw a diagram* to solve multistep multiplication problems?

PROBLEM SOLVING
Lesson 3.7

 Operations and Algebraic Thinking—4.OA.A.3 Also 4.NBT.B.5
MATHEMATICAL PRACTICES
MP1, MP2, MP4

Unlock the Problem

During the 2010 Great Backyard Bird Count, an average of 42 bald eagles were counted in each of 20 locations throughout Alaska. In 2009, an average of 32 bald eagles were counted in each of 26 locations throughout Alaska. Based on this data, how many more bald eagles were counted in 2010 than in 2009?

Use the graphic organizer to help you solve the problem.

Read the Problem

What do I need to find?

I need to find _____ bald eagles were counted in 2010 than in 2009.

What information do I need to use?

In 2010, _____ locations counted an average of

_____ bald eagles each.

In 2009 _____ locations counted an average of

_____ bald eagles each.

How will I use the information?

I can solve simpler problems.

Find the number of bald eagles counted in _____.

Find the number of bald eagles counted in _____.

Then draw a bar model to compare the _____

count to the _____ count.

Solve the Problem

- First, find the total number of bald eagles counted in 2010.

  _____ × _____

  = _____ bald eagles counted in 2010

- Next, find the total number of bald eagles counted in 2009.

  = _____ × _____

  = _____ bald eagles counted in 2009

- Last, draw a bar model. I need to subtract.

  | 840 bald eagles in 2010 |

  | 832 bald eagles in 2009 |
  ?

  840 − 832 = _____

So, there were _____ more bald eagles counted in 2010 than in 2009.

Chapter 3   183

🔑 Try Another Problem

Prescott Valley, Arizona, reported a total of 29 mourning doves in the Great Backyard Bird Count. Mesa, Arizona, reported 20 times as many mourning doves as Prescott Valley. If Chandler reported a total of 760 mourning doves, how many more mourning doves were reported in Chandler than in Mesa?

Mourning dove ▲

Read the Problem

What do I need to find?

What information do I need to use?

How will I use the information?

Solve the Problem

760 mourning doves in Chandler

580 mourning doves in Mesa

?

- Is your answer reasonable? Explain. _____

Math Talk

MATHEMATICAL PRACTICES ②

Reason Abstractly What is another way you could solve this problem?

184

Name _____

Share and Show MATH BOARD

Unlock the Problem
✓ Underline important facts.
✓ Choose a strategy.
✓ Use the Problem Solving MathBoard.

1. An average of 74 reports with bird counts were turned in each day in June. An average of 89 were turned in each day in July. How many reports were turned in for both months? (Hint: There are 30 days in June and 31 days in July.)

   First, write the problem for June.

   _____

   Next, write the problem for July.

   WRITE Math • Show Your Work

   _____

   Last, find and add the two products.

   _____ reports were turned in for both months.

2. What if an average of 98 reports were turned in each day for the month of June? How many reports were turned in for June? Describe how your answer for June would be different.

   _____

   _____

   _____

3. **GO DEEPER** There are 48 crayons in a box. There are 12 boxes in a carton. Mr. Johnson ordered 6 cartons of crayons for the school. How many crayons did he get?

   _____

4. **MATHEMATICAL PRACTICE 1** Make Sense of Problems Each of 5 bird-watchers reported seeing 15 roseate spoonbills in a day. If they each reported seeing the same number of roseate spoonbills over 14 days, how many would be reported?

Chapter 3 • Lesson 7   185

MATHEMATICAL PRACTICES   MODEL • REASON • MAKE SENSE

On Your Own

5. **THINK SMARTER** On each of Maggie's bird-watching trips, she has seen at least 24 birds. If she has taken 4 of these trips each year over the past 16 years, at least how many birds has Maggie seen?

_____

6. **MATHEMATICAL PRACTICE 1** Make Sense of Problems
There are 12 inches in a foot. In September, Mrs. Harris orders 32 feet of ribbon for the Crafts Club. In January, she orders 9 feet less. How many inches of ribbon does Mrs. Harris order? Explain how you found your answer.

_____

_____

7. **Go DEEPER** Lydia is having a party on Saturday. She decides to write a riddle on her invitations to describe her house number on Cypress Street. Use the clues to find Lydia's address.

_____

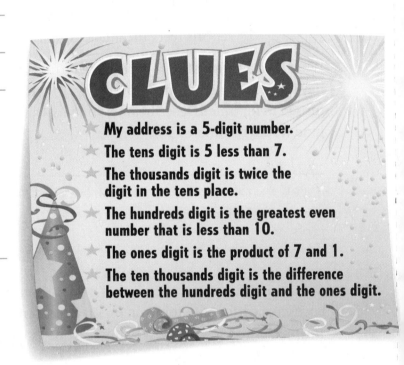

CLUES

- My address is a 5-digit number.
- The tens digit is 5 less than 7.
- The thousands digit is twice the digit in the tens place.
- The hundreds digit is the greatest even number that is less than 10.
- The ones digit is the product of 7 and 1.
- The ten thousands digit is the difference between the hundreds digit and the ones digit.

8. **THINK SMARTER +** A school is adding 4 rows of seats to the auditorium. There are 7 seats in each row. Each new seat costs $99. What is the total cost for the new seats? Show your work.

_____

Name _____

Problem Solving • Multiply 2-Digit Numbers

**Practice and Homework
Lesson 3.7**

COMMON CORE STANDARD—4.OA.A.3
Use the four operations with whole numbers to solve problems.

Solve each problem. Use a bar model to help.

1. Mason counted an average of 18 birds at his bird feeder each day for 20 days. Gloria counted an average of 21 birds at her bird feeder each day for 16 days. How many more birds did Mason count at his feeder than Gloria counted at hers?

   Birds counted by Mason: $18 \times 20 = 360$

   Birds counted by Gloria: $21 \times 16 = 336$

   Draw a bar model to compare.

   Subtract. $360 - 336 = 24$

   | 360 birds counted by Mason |
   | 336 birds counted by Gloria | ? |

   So, Mason counted __24__ more birds.

2. The 24 students in Ms. Lee's class each collected an average of 18 cans for recycling. The 21 students in Mr. Galvez's class each collected an average of 25 cans for recycling. How many more cans were collected by Mr. Galvez's class than Ms. Lee's class?

   _____

3. At East School, each of the 45 classrooms has an average of 22 students. At West School, each of the 42 classrooms has an average of 23 students. How many more students are at East School than at West School?

   _____

4. **WRITE** ▸*Math* Draw a bar model that shows how the number of hours in March compares with the number of hours in February of this year.

Chapter 3 187

Lesson Check (4.OA.A.3)

1. Ace Manufacturing ordered 17 boxes with 85 ball bearings each. They also ordered 15 boxes with 90 springs each. How many more ball bearings than springs did they order?

2. Elton hiked 16 miles each day on a 12-day hiking trip. Lola hiked 14 miles each day on her 16-day hiking trip. In all, how many more miles did Lola hike than Elton hiked?

Spiral Review (4.OA.A.2, 4.NBT.A.1, 4.NBT.A.3, 4.NBT.B.5)

3. An orchard has 24 rows of apple trees. There are 35 apple trees in each row. How many apple trees are in the orchard?

4. An amusement park reported 354,605 visitors last summer. What is this number rounded to the nearest thousand?

5. Attendance at the football game was 102,653. What is the value of the digit 6?

6. Jill's fish weighs 8 times as much as her parakeet. Together, the pets weigh 63 ounces. How much does the fish weigh?

Chapter 3 Review/Test

1. Explain how to find 40 × 50 using mental math.

2. Mrs. Traynor's class is taking a field trip to the zoo. The trip will cost $26 for each student. There are 22 students in her class.

   Part A

   Round each factor to estimate the total cost of the students' field trip.

   Part B

   Use compatible numbers to estimate the total cost of the field trip.

   Part C

   Which do you think is the better estimate? Explain.

3. For numbers 3a–3e, select Yes or No to show if the answer is correct.

   3a. $35 \times 10 = 350$ ◯ Yes ◯ No

   3b. $19 \times 20 = 380$ ◯ Yes ◯ No

   3c. $12 \times 100 = 120$ ◯ Yes ◯ No

   3d. $70 \times 100 = 7,000$ ◯ Yes ◯ No

   3e. $28 \times 30 = 2,100$ ◯ Yes ◯ No

4. There are 23 boxes of pencils in Mr. Shaw's supply cabinet. Each box contains 100 pencils. How many pencils are in the supply cabinet?

   _____ pencils

5. Which would provide a reasonable estimate for each product? Write the estimate beside the product. An estimate may be used more than once.

   | 50×20 | 25×40 | 30×30 |

   23×38 [   ] 46×18 [   ]

   31×32 [   ] 39×21 [   ]

6. There are 26 baseball teams in the league. Each team has 18 players. Write a number sentence that will provide a reasonable estimate for the number of players in the league. Explain how you found your estimate.

   [   ]

7. The model shows 48×37. Write the partial products.

   (model with 40 and 8 across the top, 30 and 7 along the side)

190

8. Jess made this model to find the product 32 × 17. Her model is incorrect.

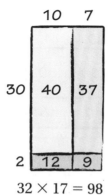

32 × 17 = 98

Part A

What did Jess do wrong?

Part B

Redraw the model so that it is correct.

Part C

What is the actual product 32 × 17?

_____

9. Tatum wants to use partial products to find 15 × 32. Write the numbers in the boxes to show 15 × 32.

(☐ × ☐) + (☐ × ☐) + (☐ × ☐) + (☐ × ☐)

10. Which product is shown by the model? Write the letter of the product on the line below the model.

    (A) 17 × 36   (B) 24 × 14   (C) 13 × 13

	10	3
10	100	30
3	30	9

	30	6
10	300	60
7	210	42

	10	4
20	200	80
4	40	16

    _____   _____   _____

11. **GO DEEPER** Mrs. Jones places 3 orders for school T-shirts. Each order has 16 boxes of shirts and each box holds 17 shirts. How many T-shirts does Mrs. Jones order? Use partial products to help you.

12. Write the unknown digits. Use each digit exactly once.

    ```
        46
      × 93
      ------
      3, □00
        5□0
        □20
    +   1□
      ------
      4,□78
    ```

    | 1 | 2 | 4 | 6 | 8 |

13. Mike has 16 baseball cards. Niko has 17 times as many baseball cards as Mike does. How many baseball cards does Niko have?

    _____ baseball cards

14. Multiply.

    36 × 28 = _____

192

Name _____

15. A farmer planted 42 rows of tomatoes with 13 plants in each row. How many tomato plants did the farmer grow?

42 × 13 = _____ tomato plants

16. Select another way to show 25 × 18. Mark all that apply.

○ (20 × 10) + (20 × 8) + (5 × 10) + (5 × 8)

○ (25 × 20) + (25 × 5) + (25 × 10) + (25 × 8)

○ (20 × 18) + (5 × 10) + (5 × 8)

○ (25 × 10) + (25 × 8)

○ (25 × 20) + (25 × 5)

17. Terrell runs 15 sprints. Each sprint is 65 meters. How many meters does Terrell run? Show your work.

18. THINK SMARTER+ There are 3 new seats in each row in a school auditorium. There are 15 rows in the auditorium. Each new seat cost $74. What is the cost for the new seats? Explain how you found your answer.

19. Ray and Ella helped move their school library to a new building. Ray packed 27 boxes with 25 books in each box. Ella packed 23 boxes with 30 books in each box. How many more books did Ella pack? Show your work.

Chapter 3 193

20. Julius and Walt are finding the product of 25 and 16.

```
   Julius        Walt
    25            25
   × 16          × 16
   150           200
  + 250           50
   500           120
               + 300
                 670
```

   Part A

   Julius' answer is incorrect. What did Julius do wrong?

   Part B

   What did Walt do wrong?

   Part C

   What is the correct product?

21. A clothing store sells 26 shirts and 22 pairs of jeans. Each item of clothing costs $32.

   Part A

   What is a reasonable estimate for the total cost of the clothing? Show or explain how you found your answer.

   Part B

   What is the exact answer for the total cost of the clothing? Show or explain how you found your answer.